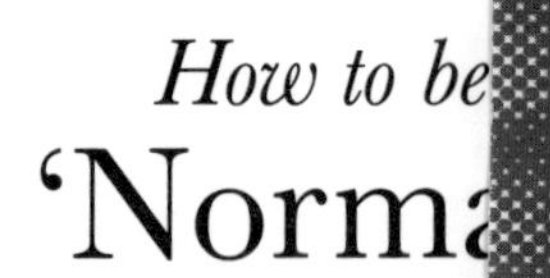

How to be

'Norma

Notes on the Eccent
of Modern Li

Formerly an autistic child whose mother tongue was numbers, Daniel Tammet is the author of a rich and widely acclaimed body of work, including memoirs, essays, literary reportage, poetry and fiction. *Born on a Blue Day* (2006) and *Every Word is a Bird We Teach to Sing* (2017) were named by *Booklist* magazine among the editors' annual selections. *Thinking in Numbers* (2012) was a BBC Radio 4 Book of the Week. Tammet's books have been translated into thirty languages. He is a Fellow of the Royal Society of Arts.

Other titles in the series:

How to Play the Piano by James Rhodes

How to Climb Everest by Kami Rita Sherpa

How to Land a Plane by Mark Vanhoenacker

How to Skim a Stone by Ralph Jones

How to Enjoy Poetry by Frank Skinner

Little Ways
to live a
Big Life

How to be 'Normal'

Notes on the Eccentricities of Modern Life

Daniel Tammet

Quercus

First published in Great Britain in 2020 by

Quercus Editions Ltd
Carmelite House
50 Victoria Embankment
London EC4Y 0DZ
An Hachette UK company

A CIP catalogue record for this book is available from the British Library.

ISBN 978 1 52941 020 4
Ebook ISBN 978 1 52941 021 1

10 9 8 7 6 5 4 3 2 1

Author photo taken by Jérôme Tabet
Illustration by Amber Anderson
Text designed and typeset by CC Book Production
Printed and bound in Great Britain by Clays Ltd, Elcograf S.p.A.

Papers used by Quercus Editions Ltd are from well-managed forests and other responsible sources.

Homo sum, humani nihil a me alienum puto.

I am a human being; nothing human can be alien to me.

Terence (*c.* 195–159 BC)

Contents

Introduction

Normal *adjective* /ˈ**nɔːməl**/ – typical, usual or ordinary; what you would expect

As a child, I spent many years learning how to be 'normal'. That is to say, how to fit in, how not to stand out, how to conform to the many social conventions I failed to grasp intuitively, the way other children did. Born on the autistic spectrum, I perceived the world around me in ways that set me apart: words had textures and colours; people resembled numbers; pi was like an epic poem.

So, I was a born outsider and, growing up, I spent many happy hours in dictionaries rather than parties. I identified with words and objects more easily than with people: I read cowboy stories not for the characters' derring-do but for the word 'saddle' with its poise and shimmer. Imagining life from a horse chestnut's, or a teacup's, or a candle flame's point of view came naturally enough to me. My classmates hardly knew what to make of me; I quickly got a name for thinking my own way, for oddness.

But like any child, I wanted to belong, and doggedly pursued 'normality' as though I were on a full-time apprenticeship.

Today, a writer and poet in my forties, I've long since come around to cherishing my difference. All those years of close and passionate observation help me now discern the glorious strangeness, often unnoticed and unsuspected, of modern life.

From tattoos and lotteries to jogging and micro-blogging, the dozen lightly humorous essays that follow are as many invitations to see our 'everyday' with new eyes. And a reminder – a lesson I long ago learned – that normality is always relative; that one person's 'normal' will be another's wondrous.

A nightclub

What chess club would be a chess club without its rooks and its bishops, its greybeards and its black-and-white boards? A book club, surely, is only ever as good as the books discussed. And if there's one valid reason to steer clear of gentlemen's clubs, it is that the members all think – so highly! – of themselves as 'gentlemen'.

How, then, to understand nightclubs? Nightclub in name only, of course. There is no night, for the bouncer won't admit the palest beam of moonlight. Nor the smallest snatch of owl song. A nocturnal chill might try to creep inside, but clubgoers have other ideas; they hurry in from the night and vanish behind the doors. They have had all evening to decide on what to wear.

The 'dress to impress' code excludes dressing gowns. And stripy pyjamas – however stripy. The young women arrive not in slippers but stilettos; the heels destined to return home hours later – courtesy of a taxi – in their hands instead of on their feet.

So, a nightclub without the night: yet clubbers pay decent money to keep it out. They pay to barricade themselves inside, in entrance fees, overpriced beer and tips. It is easy to escape daylight: think shutters or dark glasses. The night, though, is harder. You have to make a team effort: the company of many strangers to overwhelm solitude; disco balls that strew bright discs of sparkle against the dark; a DJ to break the peace of the smallest hours. Soon enough, the dance floor fills. Strobe lights fall heavy on twitching eyelids. The music thuds and thumps. The music is time counting aloud every second, like a young child rehearsing his numbers.

It is not unknown for clubbers to make up the moves as they dance. They make motions of high energy, point at random things, at the ceiling, the fire extinguisher tucked away in a corner, at the bobbing sea of faces. Some wonder how to move to be attractive, to be cool;

they are terrified of being mistaken for boring. Others think of nothing, intent on the music. Many feel free now of any need to ration their perspiration. The DJ changes his tune. The crowd throws its hundred arms up, lets down its hair, patches of brown which become red or blond when the lights play upon them. In the din of the music the dancers can't hear themselves panting. Some no longer dance but sit and watch from the bar as if off duty.

Now and again some step outside to take a 'breather', and promptly light up. The nicotine of their fifth cigarette opens their minds to the enveloping darkness. What is the night to them? Night is eight-hour workdays' culmination, an ex's shrug, a headful of unanswerable questions, their throat when it tightens.

Last call. The crowd disperses. Some ask themselves why the floor won't stay in one place. They return to their coats and, a little later, to their flats. They drop onto their beds, as asleep as asleep can be.

The tattooist

In every town and city, they do brisk business in dragons. Others specialise in life-size fruit and flowers: cherries that swell into apples on a belly once it's pregnant. Still others get a name for lurid fantasy; there's almost nothing they wouldn't tattoo. I'm talking about exploding zombies across backs and throat-slitting skeletons. (These parlours tend to have a sideline in piercings: nose, lips, right between the eyes . . .)

And then there are the tattooist's tattooists, who inscribe fine and complex geometric forms and letters, or what, to the unknowing eye, appear more or less as squiggles, as Rorschach blots of coloured ink – something, in any event, for every perplexed onlooker to ponder.

I speak as someone without tattoos, but who, a writer and poet on the spectrum, reads in each a unique and deeply personal statement. And I believe that because of this, each requires a draughtsman's skill, but also an open ear and eye. The tattooist authors words and images whose precise meanings escape him. His blank page is each client's skin, a source of self-disclosure and definition. So, before anything else, he must listen to the story his ink will write, detecting the accent of ambition, or irony, or sly timidity in which it is told. He must observe the client's clothing and the body beneath the clothing, and hear how they fit into the story. He will not begin before he understands it, at least as far as any one person can understand another person's story. He is so much more than a simple embellisher of arms, thighs and torsos.

Done well, tattooed words and images come alive along the body like the pictures and text of a well-written book in the reader's mind. The wearer of tattoos, then, becomes a breathing parchment, a Bayeux tapestry that animates itself with every gesture. Done badly, however, they have all the faults of purple prose;

they are heavy-handed, fey, unconvincing. Misspelled calligraphy is a staple of such 'purple tattooing'.

Some clients start so young that their stories might not yet be entirely stable or coherent. Dutch courage staves off any second thoughts. Like many on the spectrum, for whom the threshold for pain is low, they are hardly needle people: some would sooner faint than have their blood taken; and their epidermis cringes at the merest suggestion of a jab. Sitting in the tattooist's chair, the newbie sweats and sometimes swears, and his consciousness shrinks to a needlepoint. But the tattooist pays no attention, takes no offence – it is, he knows, only the pain talking. The tattoo occupies a long, long afternoon.

Soon, though, amazingly, the novice is good for another tattoo. And another. And a fourth. You'd think he'd be all tattooed out, but no. It proves addictive. Against the needle, an increasingly experienced arm daydreams away the pain: it is towing a young son's yellow tricycle in the garden, or sunning on a hot beach, or throwing hoops with the gang. So that initially discreet images, meant for an audience of one, now proliferate across the body. There is always more of a person's story to tell.

Small talk

It often starts with the weather: the 'nippy out', or 'rain again' or the classic 'it's turned out lovely'. Snow is always news – less so sleet or hail. The two conversers are in a café, separated by a table on which the morning's warm sun picks out the rings made by yesterday's teacups, but the women might as well be talking over a market stall, or in a hairdresser's salon or on a street corner, for the little they seem to say. To hear them speak you would never know they have twenty years' experience as mutual friends.

Small talk. A vital social lubricant. Ice-breaking patter to get our two women's conversation rolling. A swap of further pleasantries – TV shows recently watched or recorded – follows a list of the families' respective ailments: 'his hip's still playing up', 'she's a lot better now, bless her'. Broad, commiserating vowels: 'aah' and 'ooh'. Assorted compliments fill out the opening minutes of their exchange before, gradually, the language changes gears: the talk begins to grow. Particularly after downing a glass, or two, of loquacious wine.

Yet for many of those not yet north of forty, regular social contact resembles this less and less; conversation often happens far away from another person's body. A city or continent, not a table, separates them. Online, they apply themselves to typing to attract someone's attention. They type at all hours to a limit of 280 characters per message. They type things they would never say to another face: 'Hashtag tired', and 'loving Friday', and 'just gave my Xmas tie an outing. No comment.' And like the French author Georges Perec's *Attempt at an Inventory of the Liquid and Solid Foodstuffs Ingurgitated by Me in the Course of the Year Nineteen Hundred and Seventy-Four*, they keep up a running commentary on all they eat and drink: 'On my third cheese and ham sarnie of the day. V. tasty. Next up, a nice cuppa.'

To an outsider like me, the words read like the minutes kept of a life's minutiae. You have a hard time imagining the interest they might represent to other pairs of eyes, of learning 'my neighbour is mowing the grass' or 'that nap was everything'.

On the other hand, the typing won't necessarily be addressed to any one individual. You will be one of tens or hundreds, perhaps thousands, to have stumbled across these very words, most likely in a state of some distraction and half wondering whether to look away and scroll on, or pause and read. Dipping into a stranger's stream of thought, many feel the eavesdropper's perplexity. What am I missing? Yet what to them may seem like only inanities, to friends and family in the know become familiar facets of a well-loved voice, supplied by the imagination. So that these same fragments, read in the round, blur into the contours of a life story like images in a flip book.

Of course, not everything translates into words – on-screen or off. Togetherness is often best embraced in smiling silence.

Ugly vegetables

Supermarkets are theatres of head-scratching; those endless aisles make shoppers exceedingly choosy – though less so standing in Frozen Pies than in Fresh Fruit and Vegetables. Produce has a way of bringing out the perfectionist in trolley pushers, the colours inviting gazes, the shapes enthusing fingers. A melon will have its bottom sniffed. A coconut will be shaken and listened to, shell to ear, as though it held the sound of the wind soughing through the palm fronds. Such scrutiny! Anything leafy, or fiddly to peel receives even more; to within an inch of their vegetable life they are tapped, tugged and prodded.

And just when you think you have seen it all – the

cantaloupe sniffers, those with a fine ear for coconuts, the Brussels sprout botherers – you have to think again: certain fruit and veg are thrown aside, on the sole grounds that they do not look quite right. That they look ugly!

Sometimes, the ugliness runs only peel-deep. A blotchy banana, say, or a swede, good as bought, which, upon closer inspection, reveals its subtle pock marks. Nothing a light trim or a blender can't fix. But no. The banana goes untouched; the swede is tossed back. In their place, shoppers tend to stick to what they know best, cleaving to the boring basics: 'one of your five-a-day' fare, the surfaces all scrubbed, smooth and unmarked, predictably bland. In a word, cauliflower florets; in a word, King Edward spuds.

Perhaps this is why other physiognomies repel them. Take the celeriac for example, which many shoppers simply will not. True, it is not much of a looker: gnarly, lopsided, creviced. But what it lacks in looks it makes up for in a nutty, earthy sweetness. There are none so impervious to a good knife you can't reduce them to a purée.

Whenever I do my groceries, my first thought, after checking my list and subduing a celeriac-shaped craving, is to gaze around. The onion mountain rustles as hands clamber up. The air is fragrant with strawberries. There ought to be enough here for everyone, I think, from asparagus spears to zucchinis (better known under their French alias: courgettes). Yet so much food is wasted through picking only by looks. According to media reports, every second veg goes uneaten because it is considered unattractive. My other thought: in the universe of vegetables, there exist no categories like 'ugly' and 'beautiful'. In essence, fresh produce is so many calories, water, assorted vitamins and minerals: some bunched into a turnip, or tapering to a parsnip, or impersonating cabbages. All forms, cleaned and diced to pieces, are stewpot material.

A final irony bears mentioning. Other picky sniffers will purchase unsightly produce from a shop, so long as it is certified 'organic'. Carrots hairy and knobby as a grandfather's bony fingers; beat-up, snub-nosed beetroots with straggly wisps; spinach leaves caked with soil that crumbles in your palms: none of this will perturb. On the contrary, the plainer and filthier-looking the better. Naturally delicious, pesticide-free. And reassuringly ugly.

Letters

I have a friend who is a dedicated letter writer. As a correspondent she is doting, donating hours of thought each month that might just as easily have gone into a hobby. She'll put up with the paper cuts, the inky fingers, the letter writer's block that visits her intermittently. Also, the sometimes harsh, I would almost say cruel, comments from certain peers who fail to grasp her passion for longhand, and think it backward.

Such a passion for letter writing is rare these days, and especially striking when the writer is young enough to belong to the digital generation (my friend is in her thirties). Contrary to what some of her contemporaries suppose, she is in no way technophobic. She sends – the occasional, impeccably spelled and punctuated – email. She texts. She even uploads her colourful to-do notes onto a popular photo-sharing site.

But I admit to feeling daunted when I receive one of her brightly stamped letters, and worry about everything that – even as a professional writer – I tend otherwise to let hang: the occasional mangling of punctuation (an insidious email habit); vocabulary my friend might consider fit only for typing; how my handwriting slants like squally rain. Her own tight cursive hand, written with one of those humble blue biros that counsel tact and precision, becomes all her correspondence. Not for her the splay of hands over a keyboard, with every digit pitching in – left pinkie turning things plural, right thumb allotting spaces between the words. She composes one-handed, the left (like ten per cent of the population) doing all the work, walking the pen from sentence to sentence, adoring the suave flow of 'm' and 's' and 'w'. It is this that her constantly connected circle struggles to fathom. Writing's physical act matters as much to her as brushing her teeth – a need to write that is met by letters alone; the longest shopping list would be no substitute.

Sometimes I wonder if my friend's copious and exacting correspondence is an overreaction to a culture which no longer values penmanship. Is she not, by her drawerfuls of envelopes (to be addressed in the stark, impersonal capitals you otherwise see on official forms), by her hand going through the usual motions of 'dear' this and 'my very best regards' that, is she not, I ask myself, writing love letters to the letter itself, epistles to win over many others to her cause?

That sheets of white paper, wrapped in mousy paper, can inspire feelings so strong, both for and against, is extraordinary. Torn, rained-on, crumpled as a night-shirt, the letter is a modest item if ever you saw one. But to read words worth their weight in stationery is to be transported. They quicken your inner eye: you read between the lines, see the invisible crossings-out, the false starts, the half-thoughts modified or passed over: the letter as it must have read through each of its successive drafts. And you are reminded that connection is less about conveying the ins and outs of information than about sustaining friendship.

Lost property

Commuters are a people fleeced every day by their trains and buses. They are alleviated of wallets; they are separated from suitcases. It happens like this: minding well enough the gap between the train and the platform, they mind less the one that has sprung up in their attention, where the coat and hat they'd brought on with them ought to be; soon enough the clothes lie unaccompanied on the vacated seat, like the relics of some melted snowman. The same clothes that, not a quarter-hour before, they were still snugly inside.

The first they know of it is the feel of void in a trouser pocket. Or a suspicion of absence that sneaks up on their awareness. My laptop! My umbrella! Apparently, it happens all the time. Personal items race away. Many hundreds go missing daily in London alone. Quite often the passengers have form.

Is it the rocking of the carriage that lulls them into a semi-consciousness? The moving pictures of rolling landscapes? Or the ubiquity of distracting screens? A dicky memory finally conking out? Not so quick. After all, a thespian might know *King Lear* by heart, but also leave behind his script at Euston station. So, let's not second-guess anything. Whatever the truth is, the habit of losing things turns out to be fairly modern. Time was, mobility was the preserve of princes and merchants: goods went onto the back of stallions or into the belly of ships, chaperoned, escorted by troupes of vigilant attendants. Whereas for most people, belongings did not leave the house or village; precious – materially or sentimentally – possessions did not lose their way.

Sooner or later objects riding the tube or a bus on their own will be picked up. But many finders won't be keepers. Between trains themselves, or buses, always on the go, they are no more immune from shedding things along a journey; they hand in, or leave alone. So that the possessions enter the safekeeping of the Lost Property Office, an immense labyrinth of a depot that houses thousands of strangers' keys, brollies and mobile phones. Not that 'lost' is a word you would associate with quite a few of the items: the smartly gift-wrapped something; some family's good china; the proud porcelain paunch of a jug; a war medal's silver wink. Not to mention the pairs of dentures, a stuffed puffer fish, a lollipop lady's tool of trade, a judge's curly wig, the fat novel a commuter's nose had long inhabited. In short, a whole city's bric-a-brac. A hoarder's delight; a giant magpie's den.

What tales of their adventures these personal effects would tell if they could speak! As it is, their owners have three months, from the point of losing them, to make a claim. From what I can gather, reunions are as happy as they are few. Most items wind up in charity shops, or auctioned off to finance the Office. Lost clothes pursue dispersed afterlives as dusters, compost, cushion fodder.

The anti-age cream user

She has seen all the health warnings, and knows how easily she burns. Yet, come summer, she'll camouflage every inch of her pastiness beneath a tan. She'll bask away, turning limbs every quarter or half an hour, until her skin gains the rich borrowed brown of melanin.

Her eyes are bluer, her teeth whiter, since she turned mahogany; she resembles nothing so much as a chameleon that hid inside a grandmother's wardrobe. 'You look so well!' people tell her, and the compliments keep coming. 'Been on your hols, have you?' Someone even calls her 'hale'. Out and about she wears pale colours, the better to flaunt her brownness.

On a good day, in bad light, she can still pass for

forty. She is staring fifty in the face – and fifty is not smiling back. 'Give me a minute to put my face on.' Her searching eyes, with their long lashes of mascara, hold the mirror's gaze; she knows to a crow's foot every crease and wrinkle.

Adverts in the glossies make her pause: they show swipes of white cream on wrinkleless cheeks and foreheads, creams with French names and, as portrayed, aimed at the very young and flawless. She would like to be young again, she thinks. She has more or less the hang of it now. She can still turn heads with her aura of perfume. She has personality.

She takes herself to a pharmacy. Anti-age creams? There are metres of shelves of them, although the young lady sellers prefer to describe each as a 'revitalising cream' or a 'lift serum' or an 'intensive moisturising milk'. Anti-ageing, she thinks. Is anyone pro-ageing? She reads the labels, baffled; she doesn't know her AHAs from her BHAs, retinol from niacinamide, a peptide from a coenzyme. She smarts when one cream maker writes of 'giveaway signs', as if age were a guilty secret you should keep most carefully. In truth, she has nothing against her age; she likes having grown-up conversations with her children. The maker's 'giveaway signs' are – in her daughter's words – her 'laughter lines'.

But so much for laughing. She remembers the glossy ads. She doesn't want to accommodate her wrinkles; she wants them out on their ear. Silly money, really, for a pot of gloop; she buys it, all the same. She thinks, if there's any elixir inside, it is in homeopathic quantities. Simultaneously, she has the idea her skin deserves better than her; she has not treated it with enough care. Now the cream will do the looking after on her behalf. Provided the weather's not too hot. Sweating, the cream wearing off, wouldn't she age dramatically, a year every hour?

Beneath her mirror, a red wooden chest of drawers sits patiently. Paint stripper, another recent purchase, has much distressed its panels. The top and legs, as well, have been made to look far older than their years. Furniture has more 'character' that way – that's what the magazines say.

Joggers

You can hardly miss them. You see them more and more these days, out in all weathers – museum weather as much as the barbecue kind – pounding the pavements, or within the leafy city parks which are their natural habitat. I am talking, of course, about joggers.

Before being joggers, they are stretchers. They take their stretching very seriously indeed. The idea is to get their blood in the mood, warming arms and legs, making muscles – sculpted by dumbbells – comfortable. They sit on the ground and stretch. They stand and stretch. Arms extended wide describe small hoops, form upside-down Vs above the head. They hug each foot to their buttock, and viewed from the front, the joggers seem momentarily one-legged, pirates without their peg leg.

Sweatbands, pulse-taking watch, lace-up shoes – these are just some of their must-have accessories. They run on black coffee (no sugar), an empty stomach, mindless music piped in via earphones. In the name of health, they run themselves breathless. In the name of the future, they gasp and sweat week in and week out. Who knows for how many the time invested exceeds the months or years of supplementary life acquired? In the interim, many collect injuries: shin sprains, torn ligaments, runner's knee, you name it. The most sensible take time off to heal.

But when the adrenaline is coursing through their veins, drugging their blood, hardly anything can stop the veteran jogger. On occasions they backtrack to avoid a glut of tourists because the group, so dense that, as it's entered, the elbows jut out like thickets, would slow them to a halt. A park's roses might as well be plastic for all they smell them. The jogger's face is a picture of exertion. From their lungs escape big gasps of riotous breath. Blistery feet belabour the footpaths, kicking up gravel and dust, stamping their soles all over the cracks in the concrete. Studiously they keep off the grass. They take no shortcuts; always they take the long way around. Engrossed in their run, the jogger does not explore their surroundings, nor experience the perks of getting vaguely lost – the unknown spots or charming vistas discovered or the pleasure of retracing steps. They never take the path less travelled by.

They're often on their own. Their body's stamina goes to waste. Instead of running in rectangles, it could be taking them places: to the neighbouring town, say, to see a relative or a friend. But they're not always lonely. Sometimes they huff and puff in pairs, side by side, or else the fitter one makes a modest living by running two or three paces ahead; 'coach' he'll call himself. Or 'personal trainer'. To hear all their huffing and puffing you'd think the finish couldn't arrive soon enough – but there is no finish, strictly speaking – the exertion is its own end point.

Single or accompanied, they seem to run in shifts: morning, afternoon, evening, in thirty- or forty-minute slots, after which they slow to an amble, hands parked on their hips, a dazed look on their face.

Going up

Most folk prefer the easy heights of lifts. They don't know the poetry of stairs, the inner drama – not for nothing do we call the rising steps a 'flight'. It is why little children hop up them like happy sparrows, but also why, for tired legs or arthritic joints, climbing upstairs can feel a lot like piloting into crosswinds – the bumps growing ever greater until, not a second too soon, you touch down upon the landing.

For sure, poetry doesn't occupy much of a place in city lives. Lifts are everywhere these days, unsurprisingly given the vertical landscapes of office towers and apartment blocks. Grannies are their regulars, and mums complete with prams, but also brisk up-and-comers dressed for the lift: coat collar up, cap pulled down. Insomniacs so tired they rise asleep inside. And gym-hitters, on their way to an hour with the StairMaster. Some days they all happen to ride together, so that their parallel lives collide. Stomachs are sucked in; eye contact is averted; nobody speaks. Except, that is, the lift: 'doors closing'.

'Going up.' How many times in a month, a year, do the riders 'go up'? Technology elevates them constantly, without exalting their thoughts. It plays terrible tricks on their sense of perspective. A first day at the office, a skyscraper's kind. You step in the lift, hunch-shouldered and bleary-eyed, and step out a minute later, five hundred feet tall. Your new-found gigantism will last only as long as your surprise. For all the sky views afforded you, the effect soon wears off, which is perhaps just as well if you wish to rise through the ranks – literally as much as metaphorically. A giant's growing pains must be something else.

No matter the level of riding experience, lifts by their nature discombobulate. For a moment, they wrench you out of the weave of time and space, leaving your mind on the ground floor while your body exits ten floors up. Culture only adds to the confusion: American hotels routinely misname their thirteenth floor the 'fourteenth' (superstition obliging) and, by the same logic, misnumber every floor thereafter, with the result that six plus eight equals fifteen in elevator button arithmetic. Seventeen becomes not a prime number but even.

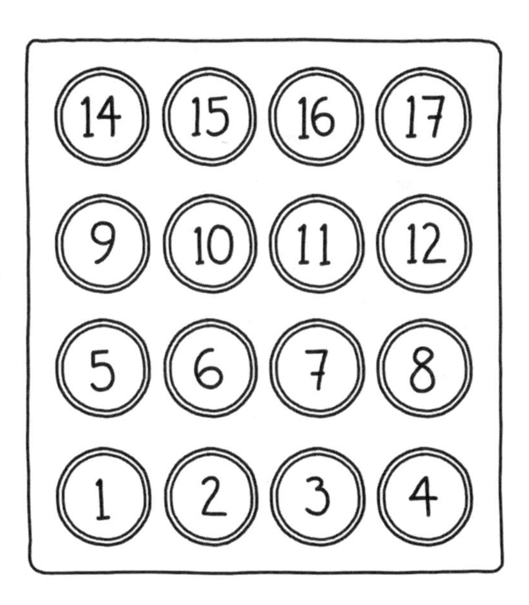

At least you know where you are with stairwells, even the fanciest ones that wind and curve. There is no more agreeable way to gain altitude; what bliss to rise and rise without the cringe of elevator music, without the crush of a lift's involuntary and surly scrum! Think of Cannes' red carpet! Think of La Scala! (which is Italian for 'stairs'). Escalators can't compete; like lifts, they flatten any sense of incremental – step-by-step – progression. A stairway is a story, a journey: twenty (or fifty, or eighty) steps in your legs, you feel you have arrived somewhere. If only at a new beginning.

Selfies

As a child, I always greeted with a groan the news that the school photographer was coming. I hated the fuss my parents made on the morning of the picture: which jumper? The reindeer or the shoulder buttons? The 'ouch!' elicited by my mother's comb as it pulled my hair. I hated the thought of the photographer's lens being trained on me. My smile was the smile of a boy told to 'say cheese'. I would squirm the whole time on the high stool I was perched on, while the prop that was a barn owl sat grumpily on my gloved and outraised hand.

But for my nieces and nephews, like most teenagers today, things could not be more different. Each is his or her very own paparazzo; they are forever taking pictures of themselves – enough photographs to fill galleries. It is so easy to do. Anytime, anywhere, they need only hold a smartphone at arm's length – not for them the tripods, rolls of film, the 'excuse me, would you mind taking our photo?' – and arrange their face before the lens, one of those faces you only ever see in photos, in the vicinity of a flash: half-frown, half-smile, smiling lips and frowning eyes.

Sometimes they down their phone and wait for a better hair day. It is most important to them, the thought of being photogenic. Always, they see and appraise themselves through other people's eyes; on social media they put their best photo forward, carefully curating their identity: here's me in front of the Eiffel Tower, and here's me enjoying a delicious brunch, and here's the lip gloss I recently bought with my best friend. They can spend an hour prepping for their 'just got out of bed' shot. It baffles me. On principle, I'd never outsource my self-image myself; I wouldn't trust random commentators with my haircut, my taste in clothes, my confidence. And yet, they do. So light, so handy, the smartphone becomes a compact mirror that remembers your every blink, every smile you approximate. Until, that is, you hit delete, and grin again and again, striking poses, pulling faces, trying on expressions galore: self-deprecation, irony lightly worn, sporting good humour. Who knew there were more than thirteen ways of looking at a camera?

A day or two away from the lens they develop withdrawal symptoms. To take selfies is as absorbing as it is permanently dissatisfying. The takers locate new backdrops, adopt ever new postures; some are quite the stretch, the likes of which a passer-by mightn't otherwise encounter in his life: postures to make a contortionist proud. Their body language in each picture says, no, that's not quite it either. They swivel their head, smile to the left and smile to the right, but always the perfect pic eludes them. What reserves of patience they possess! They wait and wait for the right fraction of a second to come along. Ever curious to see what else the camera will make of them. Ever comparing the results: which is the me that is five minutes older? They make umpteen before and after snaps: before the gym, after the gym; before the earrings, after the earrings; before the first date, after the first date. For no other apparent reason than that the moment deserves to have its photo taken.

Perhaps they hope to snap the future unawares.

Electioneering

A general election is coming to a voting booth near you. Everywhere the rumour of it goes, it brings a carnival-like atmosphere that's hard to tell from a circus. There are the many posters, which vary from town to town. Sellotaped up in certain constituencies' front windows, some advertise smart and smiling grey-haired women – like grandmothers for sale. Elsewhere, the printed leaflets carried door to door feed the kitchens' recycling bins. Neighbours discuss the party leaders as if on first-name terms with them. Polls – who's 'ahead' and who's 'behind' – become the talk of the media.

Turn on a television and you'll hear experts in polling – they are more properly called psephologists – say things like, these are good numbers for party A, or, these are bad numbers for party B, but with the proviso that in a month's time the actual results could still look very different. Solemn commentators use cryptic language like, what is the doorstep saying? And, this is how Number Ten views the situation (a curious redundancy in that 'number', as though ten might otherwise be thought a letter). Politicians in their Sunday TV best reply to questions they were not asked.

Outside the studios, the two Britains interact. There are the obligatory photo ops – the candidates donning goggles or boxing gloves, or posing with a shop till or a haddock out of water. The curious congregate around and put on a bright face, infected by the enthusiasm the politician demonstrates for, say, conveyor belts. Often, the public is outnumbered by suits with colour-coded rosettes, phalanxes of photographers, gaggles of reporters: we're live at a bread counter (outside a pub, in a bus queue . . .). Men briefly miked up become the Man on the Street; women become German noblewomen: Brenda from Bristol (Brenda von Bristol).

Whole streets are repurposed into outdoor meeting

halls; the politician steps out of a morning cab into the high noon of flashing cameras. He knocks on doors, shakes strangers' hands, kisses babies who bawl. He makes a show of rolling up his sleeves. He knows how to work a living room. Gently he taps a chair on its arm, sits smilingly at a table, plays at drinking from a teacup, hammy for the TV cameras. He mimes reassurances to people's queries as he leaves.

A drab election day morning finds many grown-ups trudging into school. They hide behind a scrawny plywood screen, as though exiled there by a teacher. Memories of the naughty corner. Sent there for growing up. On the other side of the screen, walls bright with finger paintings. Whereas on the other side of other plywood screens, up and down the country: a church's stained glass, a village pub's stools, a hair salon's blow-dryers.

I pick up a stubby pencil that grows stubbier by the hour. All around, I hear the soft crunch of paper-folding from the other booths. On the sheet I make a cross, to be shoved through a box's slot – like the first play in a game of correspondence tic-tac-toe.

The lottery

Its devotees are called players, but as games go it's no bingo. If a wag once remarked of golf that it was more or less 'televised sky', the lottery's balls hardly make better television: all suspense, zero spectacle. Even the suspense most often evaporates the instant the third or fourth ball is drawn. A measly minute or two of webcast: the narrowest of windows of opportunity to get rich. Some prefer to watch on replay, or hold back before reading the results online: why hurry, they think, there's all the time in the world to be disappointed.

No doubt the lottery novice thought his numbers luckier than the rest. He knows different now. 'Lucky' and 'unlucky' numbers are indistinguishable. The most you can ever say is, it's been a good week for round (square, cube, prime . . .) numbers – or, if you're me, a good (or bad) week for 'shy' numbers (those divisible by four), or those 'lumpy' as thirty-seven or 'loud' as fives.

Other players may be old hands at gambling: the racehorse whisperers (or so they tell themselves), the casino habitués, those who watch the roundabouts for ball bearings we call roulette wheels. But experienced or novice, their dustbins fill with the same lottery-ticket confetti. Millions, weekly, find themselves on the wrong side of fortune.

All of which comes at a cost: hope doesn't come cheap, not at two pounds the ticket. For every hundred pounds spent they'll have thirty back, on average, to show for all the hoping. A funny use of several hundred pounds per annum, not to mention the chore, the paper-work, the game becomes for them: all that biweekly number choosing and checking.

Except that many continue to swear by the same six numbers, for all the money lost on them, and are surprised when other players, taking stock, show no compunction in changing. The losses have made them wiser, these players feel, their new guesses more educated. They remain alert for possible 'patterns', keen to pit their intelligence against randomness. And if their former numbers came up? That would be their comeuppance, think the non-changers, for having turned their backs on luck after having wooed it so ardently.

Forty-five million to one: it is hard to get any head around such odds. Even mine. But the players are no one's fools; if anything, they talk down their chances. Deep inside, they know they might never win, even in a lifetime of Saturdays (and Wednesdays). And if, for many, money is often tight, and discouragement common at how expensive losing gets, they view hitting the jackpot as a long-term project. If necessary, they'll eat less, or more greasily, to subsidise their longing. To feel themselves warmed twice a week by the tingling slow release of anticipation. They have all the makings of a millionaire, they believe: the temerity for one. In the meantime, the millions they'll likely never win keep them in dreams.

Such strokes of luck belong to another time, to an only possibly attainable future – it is enough to know it lies out there somewhere before them: an immensely distant dot of light illuminating their present.